The Focke-Wulf Ta 183

David Myhra

Schiffer Military History
Atglen, PA

Published by Schiffer Publishing Ltd.
4880 Lower Valley Road
Atglen, PA 19310 USA
Phone: (610) 593-1777
FAX: (610) 593-2002
E-mail: Schifferbk@aol.com.
Visit our web site at: www.schifferbooks.com
Please write for a free catalog.
This book may be purchased from the publisher.
Please include $3.95 postage.
Try your bookstore first.

In Europe, Schiffer books are distributed by:
Bushwood Books
6 Marksbury Road
Kew Gardens
Surrey TW9 4JF
England
Phone: 44 (0)181 392-8585
FAX: 44 (0)181 392-9876
E-mail: Bushwd@aol.com.

Try your bookstore first.

The Focke-Wulf Ta 183

Allied intelligence considered this proposed turbojet-powered flying machine to be the best example of *Nazi* Germany's advances in fighter aircraft design. It came about as *Focke-Wulf Flugzeugbau's* response to *Oberst Siegfried Knemeyer*, Chief of the Technical Air Armament (*Ernst Udet's* old job) at the *RLM* (Reichsluftfahrtministerium or Government Air Ministry) call for a single *HeS 011A* turbojet powered high altitude fighter. In July, 1944 *Knemeyer* issued a Emergency Fighter Competition order for a single-turbojet-powered, high-speed/high altitude turbojet-powered

fighter especially suited to go up, engage, and bring down the American *B-29* heavy bomber. Although the *B-29* had not yet been assigned bombing duties over German cities, the *RLM* was expecting to see it any day. About mid Summer, 1944 *Heinkel-Hirth* told *Knemeyer* that their *HeS 011A* axial-flow turbojet engine was about ready for field use. But it wasn't...nor would it be prior to Germany's unconditional surrender in early May, 1945. The *HeS 011A* was Germany's second generation turbojet engine. It promised 2,866 pounds [1,300 kilograms] of thrust compared to the

first generation turbojet engines such as the *Jumo 004B* with its 1,984 pounds [900 kilograms] of thrust or *BMW's 003A* with its 1,764 pounds [800 kilograms] of thrust. More importantly, the *HeS 011A* was specifically designed to operate at high altitudes. The *Luftwaffe* had found that their first operational turbojet-powered fighter aircraft the *Me 262A-1 Schwalbe*, was seriously limited by the unreliability of the *004B* engine above 36,080 feet [11,000 meters]. This was not high enough to counter America's possible deployment of their *B-29* high altitude bomber. Thus the *Ta*

Hans Multhopp's initial *183* wind tunnel model at the *AVA*-Göttingen. *Multhopp* had supervised one of the wind tunnels there while studying for his Ph.D. in aerodynamics. After he left and joined *Focke-Wulf, Multhopp* was welcomed back at *AVA* any time to test his scale models in their wind tunnels. This was very good fortune for him but then again *AVA* was impressed by *Multhopp's* brilliance on the topic of aerodynamics.

183 was born out of Germany's desperate need in late 1944 for a heavily armed fighter by mid 1945 that could comfortably reach 47,000 feet. Competing against *Messerschmitt's Me P.1101, P.1110, P.1111; Blohm und Voss' P.212; Heinkel's P.1078C;* and *Junkers' EF.128,* the *Ta 183* was the winner and a prototype ordered about 27/28 February 1945. In about mid March, 1945 *Focke-Wulf* began work on their new contract for *Hans Multhopp's* swept-back wing fighter. A complete set of design drawings was about all that was accomplished before Germany's unconditional surrender on 7 May 1945. But that was not the end of the *Ta 183* as with so many other *RLM* proposed turbojet- powered projects. Post war the *183's* basic configuration was adopted by several European countries but especially the Soviets. The aircraft design bureau of *Mikoyan* and *Gurevich (MiG)* produced a virtual copy of the *Ta 183* and with minor modifications such as repositioning the tailplane 1/3 down on the vertical stabilizer, two fences per wing and so on they renamed the ex German *183* the *MiG 15.* Powered by a single *Rolls-Royce* "Nene" centrifugal turbojet engine of 4,080 pounds thrust the *MiG 15's* performance in the late 1940s especially in Korea during the Korean war was awesome.

The designation of this machine as "*Ta*" is misleading. "*Ta*" stands for *Kurt Tank*...a highly skilled test pilot, capable aircraft designer, and good general manager of *Focke-Wulf Flugzeugbau GmbH. Tank,* however, was very conservative and one of the last aircraft designers in Germany to accept the promising new turbojet engine as a replacement for piston motors in aircraft. He felt that the turbine's reliability was simply not yet established. He was correct, too. Unlike *Messerschmitt* and *Heinkel,* for example, who quickly embraced the turbojet engine regardless of its dismal reliability and up to three-times the fuel consumption of a high-powered piston engine, *Kurt Tank* along with *Dr.Ing. Claudius Dornier* were the last of the great German aircraft designers of *WWII* to design an airframe around the new engines from *BMW, Junkers Motoren (Jumo),* and *Heinkel-Hirth.* As a result *Focke-Wolf* and *Dornier* had only a handful of gas turbine-powered aircraft designs at the close of the war.

The father of the *Focke-Wulf Ta 183A* was a young aeronautical genius (a wiz kid he is sometimes called) by the name of *Hans Multhopp.* From time to time in aviation history publications the *183* project is referred to as *Focke-Wulf Project I.* Although this, too, is correct... throughout this book *Multhopp's* machine will be shorten to *Ta 183.* In addition, *Focke-Wulf Flugzeugbau* entered a second single turbojet engine design into the same *RLM* emergency fighter competition in mid 1944. This machine is known as the *Fw Ta 183B* or *Focke-Wulf Project II.* It was a conventional mid-wing monoplane with single-spar wing and normal-looking tail unit favored by *Kurt Tank.*

The son of a newspaperman, *Hans Multhopp* was born at Alfred, near Hannover, on 17 May 1913. Completing his general schooling in 1932 at age 19, he entered the *Technische Hochschule* (Institute of Technology) at Hannover and spent the next two years studying engineering. In 1934 he entered Göttingen University where he stayed until 1937 working toward his *Ph.D.* in the field of his burning interest...aerodynamics. It was during these years that *Multhopp* studied under the great *Dr.Ing. Ludwig Prandtl,* the father of modern aerodynamics. *Professor Prandtl* later called *Multhopp* "the most gifted student of aerodynamics that he had ever had."

While at Göttingen University, *Multhopp* designed a number of sailplanes of the general type that had become popular among college students of the day and others when civilian training in powered aircraft came under rigid regulation by the provisions of the Treaty of Versailles. *Multhopp,* along with several fellow students, built a high-performance sailplane with a wingspan of 53 feet [16.15 meters]. *Multhopp* never completed his *Ph.D.* program because he became bored with aeronautical theory and teaching and accepted a job offer from *Kurt Tank* the director of *Focke-Wulf Flugzeugbau* who promised him that he'd be able to put his advanced theories of aeronautics into practical application. While *Multhopp* was working on his *Ph.D.* program at Göttingen University he supple-

The *Ta 183* in full battle dress. Scale model by *Günter Sengfelder.*

mented his academic studies with practical experience gained through a part-time job at the *Aerodynamische Versuchsanstalt* (*AVA*) also located in Göttingen. This establishment for aeronautical research corresponded to the National Advisory Committee for Aeronautics (NACA) in the United States. Like the NACA, the *AVA* was subsidized by the government (in *AVA's* case) by the *Third Reich*, at the time through *Hermann Göring's Reichsluftfahrtministerium*. By 1937 the young *Multhopp* was in charge of one of *AVA's* wind tunnels...quite an achievement. When he accepted a job offer from *Kurt Tank* in 1938 at the age of 25 years, *Multhopp* had already published what was later to be known as one of his most important academic papers of his career...a seminal paper was on wing-lift theory. By 1940, the year of his marriage *Multhopp* had risen to be *Kurt Tank's* assistant in *Focke-Wulf's* advanced aerodynamics department. In 1943 *Multhopp* was promoted to chief designer and in 1944 was given the responsibility to design a high-altitude and high performance single-engine fighter to meet and exceed the specifications put forth by the *RLM's* Emergency Fighter Program of February, 1945. At *Focke-Wulf Multhopp's* design was given the series number *183* by the *RLM* and designated *Ta* for *Tank* in honor of his war time work at *Focke-Wulf Flugzeugbau*.

Upon acceptance by the *RLM, Multhopp's Ta 183* would be the first German jet powered fighter expressly designed for optimum performance at altitudes between 26,000 feet and 48,000 feet [7,925 meters to 14,630 meters], and with its single turbojet engine placed well aft within the fuselage. Aircraft designers were not so sure about placing a turbojet engine inside the fuselage...even by mid-1944. This was because back in 1939 *Ernst Heinkel* had placed *Hans von Ohain's* experimental *HeS 3B* centrifugal turbojet with 926 pounds thrust [420 kilograms] in his *He 178* airframe. This experimental aircraft made its maiden

flight on 24 August 1939 and several demonstration test flights later after achieving about 400 miles per hour [644 km/h] the machine was abandoned primarily because its wing lacked sufficient area for the low-thrust turbojet powering the aircraft. It would have flown nicely if powered by a strong piston engine, however, the 926 pounds of thrust from the *HeS 3B* required a wing of large surface area to help it lift off. This was the dawn of turbojet engine powered aircraft and mistakes would be made until a knowledge base was established and this knowledge base would take years. Meanwhile experts laughed at *Heinkel* and his jet-powered *He 178* which required a very long distance to get airborne... using most of the long runway. A redesign of the *He 178's* wing could have changed all that but *Heinkel* felt that it wasn't worth the effort and besides he was already concentrating on his twin turbo-jet powered *He 280* fighter prototype. But the *He 178* had left a legacy that turbojet engines mounted in the fuselage was a mistake...and no aircraft designer up to that time had placed one in side the fuselage with the single exception of the *Horten* brothers and their twin *004B Ho 9 V2* all wing fighter project. Otherwise every airframe designer placed the turbine outside...some hanging from under the wing like the *Me 262, He 280*, and the *Ar 234B* and *C* or on the fuselage like the *He 162* or the *Ju 287 V1* and *V2. Messerschmitt AG* had designed and built the *Me P.1101* in 1944 to be powered by a single *HeS 011A* mounted in the fuselage but this experimental project could not overcome its enormous technical problems stemming from its state-of-the-art variable swept-back wing mechanism to the point where it could be flown.

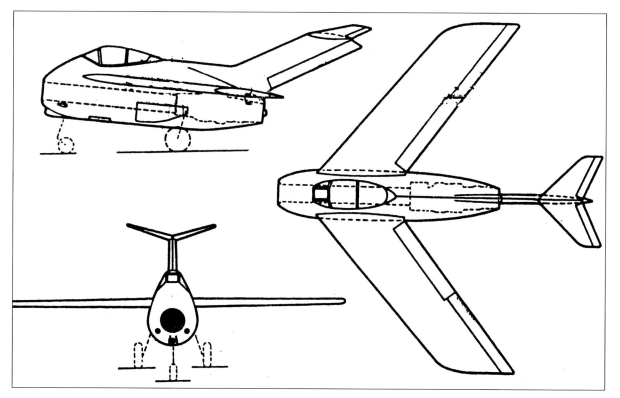

Three-view drawing of the *Ta 183*.

Hans Multhopp's Ta 183, was intended as a interceptor against high-flying Allied bombers, especially those marvelous B-29 bombers flown by the United States in the Pacific Theater during offensive military operations against Japan. Multhopp's single seat fighter, powered by the 2,866 pound[1,300 kilogram] thrust HeS 011A then under development, was expected to have a top speed of 597 mph [961 km/h] at 39,500 feet [12,040 meters] with an operational range of up to 1,150 miles [1,851 kilometers] cruising at 40% thrust. Multhopp claimed, post war, that his 183 would have had a ceiling of 47,200 feet [14,387 meters] more than high enough to intercept the high-flying piston engined B-29 American bomber. The B-29 had a service ceiling of 33,000 [10,058 meters] while the Me 262 powered by its twin Jumo 004Bs was operating at its very limits at this height. The RLM was hoping

that all this would change with the serial production of the HeS 011A turbojet engine. The HeS 011A had a thrust rating of 2,866 pounds [1,300 kilograms] compared to the 004B with its 1,984 pounds [900 kilograms] or the BMW 003A with its 1,764 pounds [800 kilograms] of thrust. Serial production of the HeS 011A would not occur prior to the war's end and only about twenty hand-built engines had been assembled. Several more were built post-war by order of the U.S. military and the new engines shipped to the United States, but in mid 1944 Heinkel-Hirth had promised the RLM that the HeS 011A was ready for serial production. With this possibility in mind, Knemeyer issued an order that all future fighters had to be designed around the HeS 011A turbojet engine. Each design was to have a top speed of at least 621 mph [999 km/h] at 23,000 feet [7,010 meters] with an operational ser-

vice ceiling of 46,000 feet [14,021 meters] attitude. Offensive fire power at this altitude had to be substantial and include four MK 108 30 mm cannon, two racks of twelve R4M unguided air-to-air rockets each, and four X-4 guided air-to-air missiles said Oberst Siegfried Knemeyer.

The design of Focke-Wulf's Ta 183 was different from any fighter aircraft the world had ever seen. It was distinguished by a "T-tail" a top of a long sweeping vertical fin...both pioneered by Multhopp and swept-back wings and tailplane. Multhopp's wind tunnel studies at his old employer as a student, AVA showed that control offered by a standard tail configuration when used in a turbojet-powered high-speed fighting machine was not good enough. His wind tunnel studies had demonstrated that during takeoff and landing, where control usually is most difficult, a very high vertical stabilizer with a hinged rudder was not affected very much by the turbulent airflow from the wings. Multhopp also found that when he placed a horizontal surface atop the vertical stabilizer/fin it gave substantially more leverage than the horizontal surfaces mounted directly on the fuselage which had been the common practice among aircraft designers.

The Ta 183 was also distinguished by having its turbojet engine installed aft in the fuselage, with the air intake and the exhaust tube running its entire length. It was further out of the ordinary by having its wings and tailplane sweptback at an angle of 40 degrees. The Ta 183's vertical tail, swept back at an angle of 60%, contained the rudder and was attached to the dorsal surface the aft fuselage. But it was the 183's horizontal stabilizer that became the proposed fighter's signature and all "T-tails" from that day forward came to be known as the "Multhopp T-Tail."

The American B-29 heavy bomber which the RLM was expecting to see in the skies over Germany in early 1945, thus the need for a high altitude single-turbojet powered fighter. The B-29 never was deployed as a bomber over Germany.

Focke-Wulf was told to develop and produce the Ta 183 at the greatest possible speed. With factories operating two shifts per day, production of the Ta 183 was expected to reach 300 per month. Initially Focke-Wulf was going to hand-build sixteen prototypes all of which were to be powered by Jumo 004Bs pending the arrival of the HeS 011As. Focke-Wulf never got beyond a full set of detailed drawings because in April, 1945 about one month after Knemeyer had awarded the contract, all their intended factories for the production of the 183 were captured by Allied troops. How soon would have the Ta 183 been ready is not clearly understood. Reflecting the shortage of war material, the Ta 183 was designed by Multhopp to be as small as possible and extremely easy to construct by using unskilled POW labor. For example, a single sheet of plywood was to be used to form the wing covering from wing root to wing tip and leading edge to trailing edge. In addition, the wings were to be interchangeable and the only metal in the wing was to be its main spar. The upper surface of the Ta 183's fuselage was to be covered by tin sheet, in fact, steel was to be used throughout the fuselage given the shortage of aluminum at this late stage of the war. Only the HeS 011A's air intake duct and vertical stabilizer were to be made out of aluminum while the horizontal tail and control surfaces were to be wood.

Hans Multhopp gave his 183, including its pilot, a large measure of armored protection. For example fuel tanks in the fuselage were protected by 15 mm steel armor plating aft and by a 3 mm steel deflector plates providing up to 10% incidence of hits along the sides and top. There was to be armor plate forward and aft of the cockpit to protect the pilot from 20 mm cannon fire. It's tricycle landing gear was to be operated hydraulically. The nose wheel would retract fully backward into the lower fuselage while the main wheels retracted forward then inward into the fuselage sides. When the main gear was fully retracted each wheel would be covered by two doors for protection, each door closing flush with the fuselage sides. The two MK 108 30 mm cannon with 120 rounds each were to be mounted in the fuselage's nose...one cannon port and one starboard the round air intake duct. There was room for two additional MK 108s with 60 rounds each, however, this would be a drastic measurer, Multhopp believed, causing the 183's climb performance, operational ceiling, and endurance to suffer substantially. Four MK 108's was not worth the loss in overall performance.

Multhopp designed his high-flying fighter to be adaptable for special uses. If it were to be used as an interceptor (designated 183-R), then it would carry a HWK bi-fuel rocket drive producing 2,205 pounds [1,000 kilograms] of thrust mounted inside the fuselage beneath its HeS 011A. This combination had already been tried on a Me 262 and with satisfactory results. The fighter interceptor HWK bi-fuel rocket drive carried T-Stuff and C-Stuff chemicals in external under wing tanks for a 3 minute burn time. Once the Ta 183-R reached its desired altitude, or when its liquid rocket propellant had been all consumed, these external tanks would be jettisoned.

Although the Ta 183 was not produced in Germany during the war as hoped, it is perhaps the most copied Luftwaffe aircraft design post war. Kurt Tank had been invited to come to Argentina by long time NAZI sympathizer President Juan Perón. The Allies had placed strict controls on those German individuals who could design and build weapons of war especially individuals like Kurt Tank. They were virtually under house arrest and could not legally leave Germany without permission of the Allies. But in 1946 Kurt Tank with the aid of the Argentines secretly left Germany via Denmark for Argentina with a roll of microfilm containing a complete set of drawings for the Ta 183. With Tank were up to sixty members of his Focke-

The Ta 183 doing what it was designed for...shooting its X-4 rocket missiles at high-flying B-29 bombers over Germany. Digital image by Mario Merino.

Wulf aircraft design group and their families including design heads *Wilhelm Bansemir, Paul Klages, Karl Thalau, Ludwig Mittlehüber,* theoreticians *Gotthold Mathias* and *Herbert Wolff,* and the noted gas dynamics specialist *Otto Pabst. Reimar Horten,* the designer/builder of the *Horten 9* twin *004B* turbojet-powered all-wing fighter joined *Tank,* too, along with former *E-Stelle* Rechlin test pilot *Otto Behrens. Horten* successfully escaped Germany via rail to Rome, Italy then the Argentine Embassy flew him to Buenos Aires. *Otto Behrens* would lose his life testing the *IAe 33.* It was in Córdoba, Argentina at the *Instituto Aerotécnica (IAe)* where the former *Focke-Wulf* engineers began work on the Argentine version of the *Ta 183* and known as the *IAe 33* "*Pulqui II*" or "*Arrow 2*." *Reimar Horten* did not participate in the *IAe 33* program except to con-

struct a 1 to 1 scale sailplane for *Kurt Tank* out of wood. *Tank* and his associates made several changes in *Multhopp's* original design. Originally *Multhopp* had mounted its wings about mid fuselage with strong fuselage bulkheads used to bolt on each wing because the turbojet engine filled up the entire internal space. Consequently the turbojet engine prevented each wing spar joining the other inside the fuselage. *Tank* argued that this arrangement with its need for substantial fuselage bulkheads to bolt the wings to because the *IAe 33* like the *Ta 183* prevented each wing spar joining each other inside the fuselage was to heavy and time consuming. Instead, *Tank's* solution was to...raise the wings higher on the fuselage...over the top of the turbojet engine...so that each wing spar would join together. Few people thought well of this

very major change. So *Tank,* in order to get some indication of the *Ta 183* with shoulder-mounted wings, allowed *Reimar Horten* to build a non- powered 1 to 1 scale aircraft to test its flight characteristics. It took *Reimar Horten* and a group of Argentine wood workers 3 1/2 months to complete the *IAe 33* sailplane. It was immediately towed into the air and test flown by *Oberleutnant Otto Behrens,* a former *Luftwaffe JG26* fighter pilot in Germany's lost Battle for Britain. Afterward *Behrens* was assigned to *E-Stelle* Rechlin as a test pilot on *Tank's* new *Fw 190* project. It has been said of *Behrens* that without him no *Fw 190* would have been ready for the *Front. Behrens'* response was that the *IAe 33's* flight characteristics were unsatisfactory. He especially didn't like the way the sailplane performed at slow speeds prior to landing. It wanted to go right into a stall during its landing approach. It also wanted to spin out in a tight turn. *Behrens* said that he was actually scared of the aircraft the way it handled itself in a cross wing all due, he believed, to its shoulder-high wing position. *Reimar Horten* modified the tail plan, putting the vertical swept-back tail fin back to *Multhopp's* original calculations, and so on. With these changes the *IAe 33* appeared to perform better, however, *Tank* would not allow *Reimar* to move the wings back to *Multhopp's* mid fuselage location nor were they changed on the *IAe 33,* either.

On 17 June 1950, about three years after *Tank* began work, the *IAe 33* made its maiden flight piloted by the *IAe's* chief test pilot *Captain Edmundo Osvald Weiss.* It was an assignment he didn't want. *Weiss* had flown *Reimar's* non-powered version of the *Pulqui II* and he knew how dangerous it was due to its tendency to stall. He asked *Reimar* what would be the best way to pilot the *powered Pulqui II? Reimar* suggested to *Weiss* that he fly the machine level at all times, make no severe banking turns, and when it came time to land touch down at a fairly high rate of speed. *Weiss* flew the *Pulqui II* just as *Reimar* suggested and the flight test went well. *President Juan Perón* and his advisors were thrilled but in private *Tank*

Oberst Siegfried Knemeyer...Chief of Technical Development of the *RLM* who issued a request for single turbojet-engine *HeS 011A* powered fighters early to mid February, 1945.

was furious with *Weiss* for gently piloting the *IAe 33* the way some old "woman" would. *Weiss* was never asked to fly the *IAe 33* again nor did he want to.

Otto Behrens was the next person to fly the *IAe 33*. Bringing the *IAe 33* down to land the aircraft bounced three times damaging its tricycle landing gear. Again *Tank* was furious but so was *Behrens*. He told *Tank* that in all of his years as a fighter pilot with *JG26* and even more years as a test pilot with *E-Stelle* Rechlin, that the *IAe 33* was a "brute of an aircraft" and the most dangerous machine he'd ever flown...ever flown! Several days later the *IAe 33's* damaged landing gear had been repaired and its oleo leg struts strengthened. *Kurt Tank* decided that he'd take the machine up for its third test flight in October, 1950. Upon landing *Tank* bounced the *IAe 33* just as *Behrens* had done before him but this time the gear

wasn't damaged. *Tank* again flew the *IAe 33* for its first official public demonstration flight. This was in Buenos Aires on 8 February 1951. *President Perón* and many of Argentina's highest officials were present. Immediately after takeoff, *Tank* climbed steeply to 7,500 feet [2,286 meters]. He then went into a dive, leveled out at 300 feet [91 meters] in front of the spectators, and sped across the airfield with throttle full open at more than 600 mph [966 km/h]. Turning the *Pulqui II* around, he then headed back over the airfield, this time 120 feet [37 meters]. *Tank* then returned once more over the airfield, only this time at a mere 75 feet [23 meters] above the wildly cheering crowd.

The *IAe 33 Pulqui II* was never put into production and only five of the machines were completed. The last flight of a *IAe 33* was on 18 September 1959. Although the design was impressive and modern

enough with a superb *Rolls-Royce* "*Nene II*" centrifugal turbojet engine of 5,088 pounds [2,308 kilograms] thrust, test flights had shown that *Tank* had turned the aircraft into a dangerous aircraft which pilots declined to fly...a real design failure. Two men had already died test flying the *IAe 33* prototypes. *Tank* had not experienced the death of a test pilot in one of his designs since popular aerobatic pilot *Paul Bäumer*, the recipient of the *Pour le Mérite* for his flying in *WWI* had died in the crash of *Rohrbach Metall Flugzeugbau's* "*Rofix*" prototype in 1927. A German court of inquiry later blamed *Tank* for building an aerobatic aircraft as a monoplane [single wing] when it should have been a bi-wing. It appeared that neither love nor money could never entice *Multhopp* to come to Argentine and work with *Tank* on *IAe 33*. *Multhopp* at this time was working in the United States for the *Glenn Martin Aircraft* and he could have taken a leave of absence but the

A virtual copy of *Multhopp's Ta 183* right down to nearly every nut, screw, and bolt...the awesome *MiG 15*. The machine pictured here was flown to the American Kimpo Air Base in Korea on 21 September 1953 by North Korean pilot *RO Kim Suk*. The *MiG 15* was then dismantled, transported to a U.S. Air Force installation on Okinawa, and reassembled for tests by American technical experts. Official U.S. Air Force photo taken at Wright-Patterson Air Force Base, Dayton, Ohio.

two men had never been friends. *Multhopp* was considered by his colleagues as a brash highly intelligent wiz kid who followed the dictates of his wing tunnel research while *Tank* was a practical conservative person who looked to past experience to guide him. *Tank* believed that one should not be the first when the new are tried nor be the last when the old is cast aside. Two design engineers of aviation yet with personalities and attitudes worlds apart. With out the design genius of *Hans Multhopp* in Argentina to help *Tank* and his staff build an Argentine version of the *Ta 183*, alias the *Pulqui II*, *Tank* messed up the aircraft's wing-lift aerodynamics giving the machine "superstall" problems at slow speed, especially during landing. Two good men had lost their lives test flying *Tank's IAe 33*: former *Oberleutnant Otto Behrens* and Argentine pilot *Captain Manual*. The *V1* was lost due to inadequate welding in the wing, and another *Pulqui II* prototype crashed in front of *President Perón* and guests during an air show. For a number of reasons, *Kurt Tank* and all of his former *Focke-Wulf* colleagues were fired by the Argentines in 1953 and told to make preparations to leave. The *Pulqui II* was canceled after only five prototypes. Instead, the Argentine's found that they could purchase surplus American-made *F-86's* for the price of one *Rolls-Royce* "*Nene*" turbojet engine. The Argentines chose to purchase 100 used *F-86's* from the United States in 1954.

On the other hand, the Soviet's had found a second roll of microfilm containing engineering drawings for the *Ta 183* when they sacked the *RLM*-Berlin in March, 1945. Their version of the *183* known as the *MiG 15* made its maiden flight on 2 July 1947 powered by a single *Rolls-Royce* "*Nene*" centrifugal turbojet engine with 4,850 pounds [2,200 kilograms] of thrust. *Rolls-Royce* had earlier sold twenty-five "*Nenes*" to the Soviet Union in September, 1946. The Soviets, after taking the engine apart and studying its design found several ways to increase its thrust up to 5,952 pounds [2,700 kilograms] and called it the *VK-1* after *Major-General Vladimir Klimov*. To the astonishment of *Rolls-Royce* engineers back in England the *Klimov* team in August, 1947 found that with afterburning the old "*Nene*" now known as the *VK-1F* was producing 7,450 pounds [3,380 kilograms] of thrust! The Soviet aircraft design team of *Artem I. Mikoyan* and *I. Gurevich* (known as *MiG*) found that *Multhopp's* swept-back wings permitted higher maximum forward speeds, but they also tended to produce critical stalling characteristics at low forward air speeds. This occurred especially when the airflow over the wing tended to "wash out" thus stalling spanwise across the wing rather than flow straight back from leading edge to trailing edge. For this reason, the *MiG* team reported, the *Ta 183* was difficult to maneuver at slow-speed. Part of the problem, they believed, was that the angle of the wing to fuselage was not correct although its mid-fuselage placement by *Multhopp* was correct. They discovered that only a few degrees of anhedral was needed to replace the dihedral of the *Ta 183* prototype. The solution to the "wash out" or low-speed stalling required either the use of automatic opening and closing apertures on the wing's leading edge (also known as leading-edge slots) to provide the necessary upward force at low-speed to keep the air stream flowing smoothly back over the wing's surface. Or, said the *MiG* team, a cheaper but equally effective method was the use of vertical surfaces known as "boundary layer fences" on the upper surface of the wing to reduce the outward movement of the air stream during landing. *MiG* chose to go for two fences per wing. Now *Multhopp's Ta 183* landing approach could be as slow as 109 mph [175 km/h] with very little difficulty.

Finally the design team of *MiG* took a look at *Multhopp's* T-tail on the *183*. Wind tunnel studies carried out by the Soviets indicated that the tailplane which *Multhopp* had placed on top of the swept-back vertical fin should be lowered about 1/3 down from *Multhopp's* original design. The *MiG* team agreed with *Multhopp* (we have to realize that *MiG's* wind tunnel observations were based on four to five-year's worth of new aerodynamic understanding/study) that for high-speed flying the tailplane (also known as the horizontal stabilizer) was best mounted high above the wing in order to prevent wing washout from inter-

The *MiG* design team....A. Mikoyan (middle) and M. Gurevich (left).

fering with its effectiveness. However, for low-speed flying, the horizontal stabilizer at low angles of attack (when the machine is landing for example), the optimum stabilizer position should be as lower on the vertical stabilizer as possible for the same reason. The *MiG* team found that *Multhopp's* T-tail on the *183* required a compromise setting for practical reasons, that is, during landing the T-tail tended to stall. To avoid this undesirable characteristic pilots were forced into a touch proposition requiring the need of high landing speeds. The *MiG* team felt that when the horizontal stabilizer was set about 1/3 down from *Multhopp's* initial position slower landing speeds were obtainable. With these changes in mind, the *MiG* designers modified *Multhopp's Ta 183.* The prototype made its maiden flight on 30 December 1947 and its flying characteristics were found to be satisfactory. On the basis numerous flight tests the modified *183* was ordered into serial production in March, 1948 and designated the *MiG 15.*

The Soviet's *MiG 15* was considered roughly comparable in both mission and in performance to the

The characteristic swept back wing and tail assembly signifying that a *MiG 15* is overhead.

USAF's *F-86A* "*Sabre*" and where they met in combat over Korea, performance wise, the *MiG* and the *F-86A* were generally comparable, too. The *MiG* being slightly smaller and lighter than the *F-86A*, was a better aircraft above 30,000 feet [9,144 meters]. It had a slightly higher top speed and higher rates of climb and zoom than the *F-86A*, at these altitudes. On the other hand, the *F-86A* was basically a cleaner machine than the *MiG* and, therefore, less susceptible to high-speed drag-raising effects, and was able to out-maneuver the *MiG* in all dives. The *MiG* could maintain its rate-of-climb advantage, while the *F-86A* had the ability to turn more sharply, at all altitudes. All in all, the two machines were very nearly equal in capability, but the greater success attained by the *F-86A*'s in Korea can be attributed principally to the superiority of American fighter pilot training.

With Germany's surrender *Hans Multhopp* received numerous offers from Allied countries anxious to design and build turbojet-powered aircraft. He accepted an offer from the *Royal Aircraft Establishment* (*RAE*) Farnborough, Hants, England. *Multhopp* produced a design for a high-speed research aircraft capable of reaching 800 mph [1,287 km/h] in level flight. It was turbojet-powered and having a wing sweep of 55 degrees. The pilot was to lie prone beneath the turbine's air intake duct leading to the *Rolls-Royce AJ65* "*Avon*," their first axial-flow turbojet engine. The proposed experimental aircraft would have no wheeled landing gear but would have used jettisonable wheels for takeoff and a pair of built-in skids during landings. Landing speed was estimated to be 160 mph [257 km/h]. By British standards, *Multhopp's RAE* design was pretty radical stuff, with its mid-mounted 25 foot

long [7.6 meters] wing sweptback 55° at the quarter chord. The tube-like fuselage was 26.6 feet [8.1 meters] long and had a diameter of 3.3 feet [1 meter], and the tail, which was swept 64.5°, carried the horizontal surface at its top...a *Multhopp* trademark. This *RAE* supersonic research aircraft was destined never to be built due mainly to the worsening economic conditions throughout England post war. However in 1947 the British aircraft firm of *English Electric* did construct a prototype *Mach 2.0* interceptor they called the "*Lighting.*" It became the world's first truly operational supersonic aircraft, built on the basis of official specifications that grew out of the design conceived by *Multhopp* for the supersonic *RAE* project and later strongly articulated by him in his seminal theory on lift surface for swept wings.

In 1949 *Multhopp* left England and accepted a position with the *Glenn L. Martin Aircraft Company*, Baltimore, Maryland. There he participated in the design and construction of the *Martin XB-1* bomber, the *P5M-2* "*Seaplane*," and the *XP6M* "*Seamaster*." Post war *Multhopp* was not happy professionally because his work in England and later in the United States never approached those exciting times in 1943-1945 when he was working on his *183* interceptor project. His work on the *Martin* "*Dyna Soar*" a lifting body type of space shuttle re-entry vehicle brought him a small measure of the personal satisfaction he longed for. In 1967 *Multhopp* joined *General Electric Company's* Space Re-entry and Environmental Systems Department. *Hans Multhopp*, the most gifted aerodynamicist *Professor Ludwig Prandtl* had ever taught, holder of aircraft patents, numerous acclaimed papers, a design genius, and a true lone wolf, died on 30 October 1972 at Cincinnati, Ohio at the untimely age of 59 years.

Ta 183 (*standard fighter*) *Specifications*:

- Type - Fighter
- Country - Germany
- Manufacturer - *Focke-Wulf Flugzeugbau*, Bremen
- Designer- *Hans Multhopp*
- Year Constructed - Designed mid 1944 but not con - structed prior to war's end
- Power Plant - *1xHeinkel-Hirth HeS 011A* axial-flow tur- bojet engine having 2,866 pounds [1,300 kilograms] of thrust
- Wing Span - 32.8 feet [10 meters]
- Wing Area - 242 square feet [22.5 square meters]
- Wing Loading At Maximum Flying Weight - 46.4 pounds per square inch
- Aspect Ratio - 4.45
- Length - 20.6 feet [6.3 meters]
- Height - na
- Weight, Empty - 6,570 pounds [2,980 kilograms]
- Weight, Takeoff - 9,480 pounds [4,300 kilograms]
- Weight, Maximum Flying - 11,200 pounds [5,080 kilograms]
- Fuel, Internal - 2,646 pounds [1,200 kilograms] or about 336 gallons
- Crew - 1
- Speed, Maximum - 543 mph at 23,000 feet and 597 mph at 39,500 feet [925 km/h at 7,010 meters and 960 km/ h at 12,040 meters]
- Speed, Cruise - 391 mph [629 km/h] at about 60% throttle opening
- Speed, Landing - 103 mph [165 km/h]
- Take-off Run - na
- Range, Maximum - 348 miles [560 km] at 100% throttle opening at sea level
- Flight Duration - 30 minutes at 100% throttle opening with internal fuel tanks of 2,646 pounds [1,200 kg] or 4.5 hours at 40% throttle opening with internal and external fuel tanks of 4,409 pounds [2,000 kg]
- Ceiling - 47,200 feet [14,000 meters]
- Rate of Climb - 79.5 feet/second [24.2 meters/second] at sea level or 53.5 feet/second [15.3 meters/second] at 19,700 feet [6,000 meters]
- Armament - 2x*MK 108* 30 mm cannon with 120 rounds each, 2x*R4M* racks of 12 unguided air- to-air rockets, and 4x*X-4* guided air-to-air missiles
- Bomb Load - 1x1,102 pounds [500 kilogram] bomb or torpedo

A *Ta 183* sort of look alike...a stack of Swedish *SAAB J-29's* in the early 1950s.

IAe 33 Pulqui IIa

- Type - Fighter
- Country - Argentina
- Manufacturer - *Instituto Aerotécnica* (*Fabrica Militar de Aviones*)
- Designer - *Hans Multhopp's Ta 183* design modified by *Kurt Tank*
- Year Constructed - 1947
- Power Plant - *1xRolls-Royce* "*Nene II*" centrifugal turbo-jet engine having 5,088 pounds [2,308 kilograms] of thrust
- Wing Span -34.8 feet [10.62 meters]
- Wing Area - 270 square feet [25.10 square meters]
- Wing Loading At Maximum Flying Weight - na
- Aspect Ratio - na
- Length -38.1 feet [11.6 meters]
- Height - 11 feet [3.35 meters]
- Weight, Empty -7,835 pounds [3,554 kilograms]
- Weight, Takeoff - na
- Weight, Maximum Takeoff -13,201 pounds [5,988 kilograms]
- Fuel, Internal -4,134 pounds or about 500 gallons [1,875 kilograms]
- Crew - 1
- Speed, Maximum - 646 mph at 15,748 feet [1,040 km/h at 4,800 meters]
- Speed, Cruise - 597 mph at 26,246 feet [960 km/h at 8,000 meters]
- Speed, Landing - 110 mph [178 km/h]
- Take-off Run - 2,428 feet [740 meters]
- Range, Maximum - 1,261 miles at 32,808 feet [2,030 kilometers at 10,000 meters]
- Flight Duration - 1 1/2 hours
- Ceiling - 49,212 feet [15,000 meters]
- Rate of Climb - 98.5 feet/second [30 meters/second]
- Armament - 4xOerlikon 20 mm cannon
- Bomb Load - 1,023 pounds [464 kilograms]

One of the *Ta 183's* competitors...the *Junkers Ju EF 128.* It was a tailless design with an all wood shoulder high wing. Lateral control surfaces were mounted above and below the wing just inboard the ailerons. Air intakes for the single *HeS 011A* were on each side of the fuselage under the wing root. *Junkers'* engineers went so far as to build a full-scale mockup after wind tunnel tests were highly satisfactory. Two fuel tanks were located in the fuselage containing a total of 477 gallons. Scale model by *Dan Johnson.*

Left: Another *Ta 183* competitor...the *Blohm & Voss Bv P.212.03*. It was a tailless design with swept back inverted gull wings. Lateral control surfaces were near the tip only on the upper side. The control surface on the downward sloping wing tip served as an elevator and also partially fulfilled the function of a rudder and aileron. The wings were to be constructed of steel and contained one fuel tank. A third fuel tank was located in the fuselage and the *P.212* carried a total of 330 gallons. Air was fed to the single *HeS 011A* via an open fuselage nose. Three *MK 103 30 mm* cannon were mounted in the fuselage nose. Scale model by *Dan Johnson*.

Center: The *Heinkel He P.1078C* also went up against the *Ta 183*. It was a true tailless machine with a pronounced inverted gull wing which carried all of the aircraft's 320 gallons of fuel. The air intake was located in the nose and was flattened to provide space for the pilot's cockpit. Water color by *Loretta Dovell*.

Right: The *Messerschmitt Me P.1112*. This *Messerschmitt* V-tailed competitor to the *Ta 183* suffered too many design problems to be much of a challenger. Shown is a wind tunnel scale model. The wing design came the *P.1101* and *Messerschmitt* engineers sought to achieve the smallest diameter cross section with their *P.1112*. This design is characterized by its V-tail. Self-sealing fuel tanks held a total of 265 gallons in the fuselage and an additional 44 gallons in a tank beneath the *HeS 011A* turbojet unit. Air intakes were placed aft on the fuselage sides above the wing root and just forward their leading edge. Initially the *P.1112* had been designed with a conventional tail but it was later changed to the V tail and entered into the *RLM's* single-engine competition.

This project was one of *Professor Messerschmitt's* favorite projects to prove the soundness of the swept-back wing. It was also a design that simply failed to jell as its designer *Woldemar Voigt* struggled for over a year to perfect. A prototype was constructed and was captured intact at Oberammergau southwest of Munich by American troops. Postwar *Bell Aircraft* built two prototypes patterned after the *P.1101* designated the *X-5*. The prototypes after being tested were rejected by the American Navy.

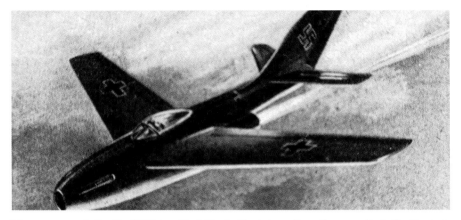

The proposed *Fw Ta 183A* by *Kurt Tank*. *Tank* wasn't quite sure that *Multhopp's* exotic T- tailed single engined fighter would make it pass *RLM's* judges about 27/28 February 1945 looking for a high altitude, single engine jet fighter to bring down the high flying American *B-29* bomber. *Tank* entered this design, too, to better his odds of winning the competition. In the end it was *Tank's* rather conventional-looking machine which failed to impress the judges and was ignored.

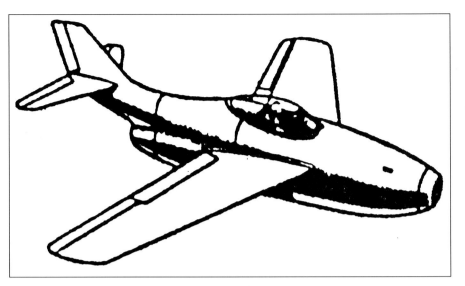

A pen and ink drawing of the proposed *Fw Ta 183A*.

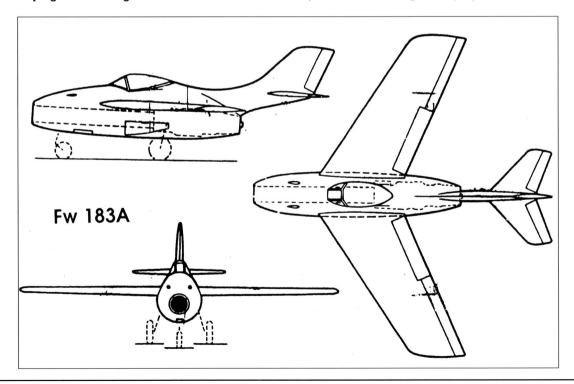

Fw 183A

A pen and ink three-view drawing of the proposed *Fw Ta 183A*. In several ways this design approximates the changes *MiG* made to *Multhopp's Ta 183B* turning it into the *MiG 15*.

The *Heinkel He 178B.* The first air frame designed for air flow through the nose and exiting out the fuselage tail. A nice design, however, *Heinkel* underestimated the wing area (square footage) required for the early turbojet engines...it was too low and as a result the *He 178B* performed poorly.

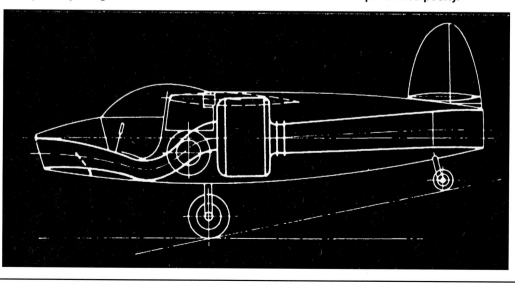

A pen and ink drawing of the *He 178A* showing the air intake for its *HeS 03* centrifugal turbojet engine and its very long tail pipe.

Kurt Tank (right) after a test flight in his *Fw 190* sharing a humorous private moment.

Kurt Tank (right) with the 5th highest scoring *Luftwaffe* ace *of WWII (*258 confirmed kills*) Major Walter Nowotny* looking over a scale model of *Tank's Fw 190* fighter.

A starboard side view of *Kurt Tank's Fw 190.*

A starboard side view of *Kurt Tank's* long-nose *Fw 190D* shown post-war at Freeman Field, Ohio.

Kurt Tank's "*Rofix*" from *Rohrbach Metall Flugzeugbau GmbH* in which *Paul Bäumer* crashed into the North Sea and died.

Paul Bäumer and famed German female aerobatic pilot *Thea Rasche*. Hamburg about 1926.

Professor Dr.Ing. Ludwig Prandtl of Göttingen University (center) said that *Hans Multhopp* was the most gifted student of aerodynamics he'd ever taught. The two *Luftwaffe* officers shown are the *Horten* brothers...*Walter* (left) and *Reimar* (right). The two civilians are *Prandtl's* aids. About 1944.

Hans Multhopp (right) chatting with noted German physicist *Professor Adolf Becker*...both members of the German Academy for Aeronautical Sciences. *Multhopp* was the Academy's youngest member ever. Bad Eilson. About Fall, 1944.

Hans Multhopp holding a scale model of his *Ta 183* high-altitude fighter. *Glenn Martin Aircraft*, Baltimore, Maryland. About 1954.

The *Multhopp Ta 183* shown in camouflage paint with its *HeS 011A* burning at full thrust. Digital image by *Mario Merino*.

One of *Multhopp's 183s* in camouflage chasing down a British "*Spitfire.*" Digital image by *Mario Merino*.

Direct starboard side view of a dual camouflaged *Ta 183* carrying two *X-4* air-to-air guided missiles under its starboard wing. It appears to be waiting for take-off clearance. Digital image by *Mario Merino*.

A *Ta 183* viewed from its rear port side and appearing armed to the teeth with four *X-4* air-to- air guided missiles for use against *B-29* formations over Germany. Digital image by *Mario Merino*.

A nose, port side view of a *Ta 183* with out its pilot. The fighter has its full complement of *X-4s* mounted under its wings. Digital image by *Mario Merino*.

A *Ta 183* armed with a different type of offensive armament: a rack of under the wing mounted unguided *R4M* "*Orkan*" rockets. These items were effective against bomber packs. Digital image by *Andreas Otte*.

A factory fresh-appearing *Ta 183* with only its four *MK 108* cannon. Scale model by *Günter Sengfelder*.

A full port side view of *Hans Multhopp's Ta 183* in camouflage. Scale model by *Dan Johnson*.

A *Ta 183* armed with four *X-4* guided missiles posed, ready, and waiting for takeoff instructions. Digital image by *Mario Merino*.

A *Ta 183* in camouflage paint appearing as if it is beginning its take-off roll. Digital image by *Mario Merino*.

A *Ta 183* at full take-off thrust its *HeS 011A* single turbojet engine throwing a flame well out beyond the fighter is pushing the machine ever faster down the runway somewhere in Germany in search of *B-29 superfortress'* seen on *Luftwaffe* radar. Digital image by *Mario Merino*.

One *Ta 183* lifting off in search of American *B-29* high altitude bombers with its *X-4* guided missiles ready for action. Digital image by *Mario Merino*.

A direct overhead view of a *Ta 183* with its four *X-4* guided missiles extending out from under the high altitude fighter's wing leading edges. Digital image by *Mario Merino*.

Two high altitude *Ta 183s.* The upper *183* has spent its four *X-4s* while the *183* in the foreground is reading itself for an attack. Digital image by *Mario Merino*.

A direct frontal view of a *Ta 183* on the starting line. Digital image by *Mario Merino*.

A pair of *Ta 183s* parked out front of their hangar after returning from a mission and shown with their *X-4s* gone...most likely fired earlier at intruding American *B-29* bombers. Digital image by *Mario Merino*.

A low altitude-radar avoiding offensive mission into England via the White Cliffs of Dover: A *Heinkel He 343* escorted by a *Ta 183*. Composite image by *Dan Johnson*.

A full front view of a *Ta 183*. Scale model by *Dan Johnson*.

A rear starboard side view of a camouflaged *Ta 183*. Scale model by *Mark Hernandez* and photographed by *Tom Trankle*.

A nice overhead view of the *Ta 183*. Scale model by *Mark Hernandez* and photographed by *Tom Trankle*.

A full front view of a *Ta 183* and showing its straight-through air intake. Scale model by *Mark Hernandez* and photographed by *Tom Trankle*.

Rear port side view of a *Ta 183* showing the exhaust duct of the *HeS 011A*. Scale model by *Mark Hernandez* and photographed by *Tom Trankle*.

A starboard side front view of a *Ta 183* and giving a good view of its tricycle landing gear doors. Scale model by *Mark Hernandez* and photographed by *Tom Trankle*.

A direct port side view of a *Ta 183* in camouflage paint. Scale model by *Mark Hernandez* and photographed by *Tom Trankle*.

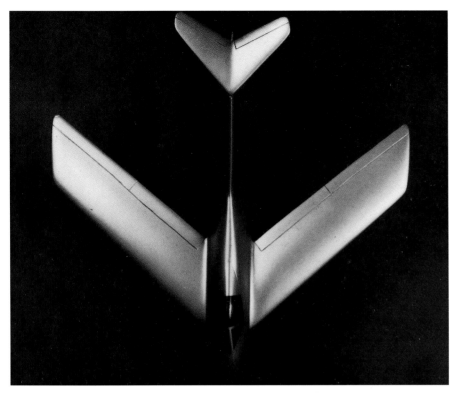

The 2ⁿᵈ original *183* scale wind tunnel model used by *Hans Multhopp* at *AVA*-Göttingen. The *183* is seen from above. About late 1944.

Multhopp's 2ⁿᵈ *183* scale wind tunnel model as seen from below. The item found mid fuselage is a fixture for positioning the model in the mouth of the wind tunnel. About late 1944.

Multhopp's 2nd *Ta 183* scale wind tunnel model resting on its *HeS 011A* exhaust duct. Shown is the model from its front starboard side. About late 1944.

Multhopp's 2nd *Ta 183* scale wind tunnel model as seen from the port rear side. About late 1944.

A direct front on view of *Hans Multhopp's* 2nd *Ta 183* scale wind tunnel model. The round rod protruding from under the air intake duct is used to fix the model in the wind tunnel's mouth. About late 1944.

A direct rear view of *Hans Multhopp's 2ⁿᵈ Ta 183* scale wind tunnel model. About late 1944.

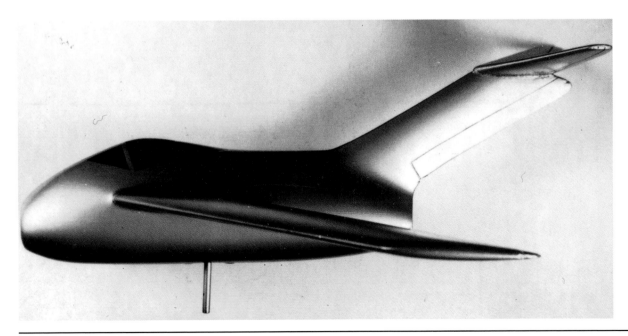

A direct port side view of *Hans Multhopp's 2ⁿᵈ Ta 183* scale wind tunnel model. Its long sweeping vertical stabilizer with its attached rudder is featured. About late 1944.

The wind tunnel at *AVA* where *Hans Multhopp* wind tunnel tested the shape of his 1ˢᵗ *183* prototype fighter. Shown is the 1ˢᵗ *Ta 183* scale model fixed in the mouth of one of *AVA's* wind tunnels at Göttingen and identified by its downward-bending tailplane. About mid 1944.

A collection of *Focke-Wulf Flugzeugbau*-related designers. Left to right: *Käther*, two unknown *Luftwaffe* officers, *Stampa* the builder of the *Ta 183* paper and wood 1ˢᵗ scale model version, *Mittlehüber*, *Naumann*, and *Otto Pabst*. These men are shown on the steps of the spa-hotel at Bad Eilson where *Focke-Wulf* design offices from Bremen had been relocated after they had been destroyed by American *B-17* heavy bombers. It was said of *Stampa* that he could build a paper and wood model shown in this photo in as little as one day's time. Summer, 1944.

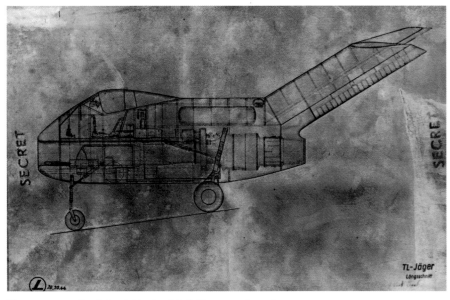

A complete, ready to run *HeS 011A* as seen from its port side. This 2nd generation axial flow turbojet was rated at 2,866 pounds [1,300 *kg*] thrust and would have powered most of the Luftwaffe's high altitude fighters.

Photo copy of original *Focke-Wulf Flugzeugbau* drawings from early 1945 showing the proposed *Ta 183's* port side internal components.

Photo copy of original *Focke-Wulf Flugzeugbau* drawings from early 1945 and featuring the *Ta 183's* wing fuel tanks for its *HeS 011A* turbojet engine. Another fuel tank was located in the aft fuselage and can be seen above the *HeS 011A*. Total amount of fuel carried internally was approximately 336 gallons. The *Ta 183* is seen from its front port side.

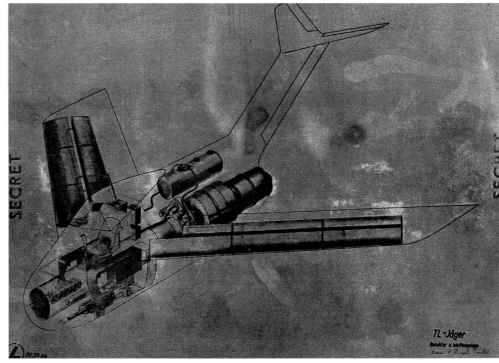

A second version of the *Ta 183* (known as the *183R*) in addition to its *HeS 011A* would have had a auxiliary *HWK* bi-fuel liquid rocket engine mounted for use in very rapid climbing ability above its turbojet engine but attached to it. Thrust from this auxiliary *HWK* would have exited via the long pipe under the fuselage. This special unit would have produced 3,750 pounds of thrust for about 3 minutes. Shown in this photo is a *Me 262C1-A* with a *HWK* auxiliary bi-fuel liquid rocket drive. This *Me 262C1-A* could reach 26,250 feet altitude upon takeoff in 4 1/2 minutes. *C Stoff* and *T Stoff* would have been carried in wing-mounted tanks. Scale model by *Jamie Davies*.

Initially the production *Ta 183's* were to used *Jumo 004Bs* up until the time the *HeS 011A* had entered serial production. The *011A* would not reach serial production although about twenty units had been made available for field testing. Shown here is the *Jumo 004A* which is the same as the *004B* except some of the exterior plumbing and a greater amount of scarce metals.

The series production *Ta 183* was designed to carry up to four *MK 108* 30 *mm* cannon. Shown here with *Flight Lieutenant*, later first rate aviation historian, *Alfred Price*, is one *MK 108* 30 mm cannon of the type to be carried aboard the *183*. England about late 1945.

A partial see-through pen and ink illustration of the *MiG 15*. In its fuselage (aft) is the Soviet version of the *Rolls-Royce* "*Nene*" centrifugal turbojet engine.

This is how the *X-4* air-to-air guided missile was suppose to work. Several *Ta 183s* would fire into the *B-29* pack and speed away allowing other *Luftwaffe* fighters such as *Fw 190s* and *Bf 109s to* pick off the damaged but still flying *stragglers*. Courtesy of *German Aircraft Industry and Production 1933-1945*.

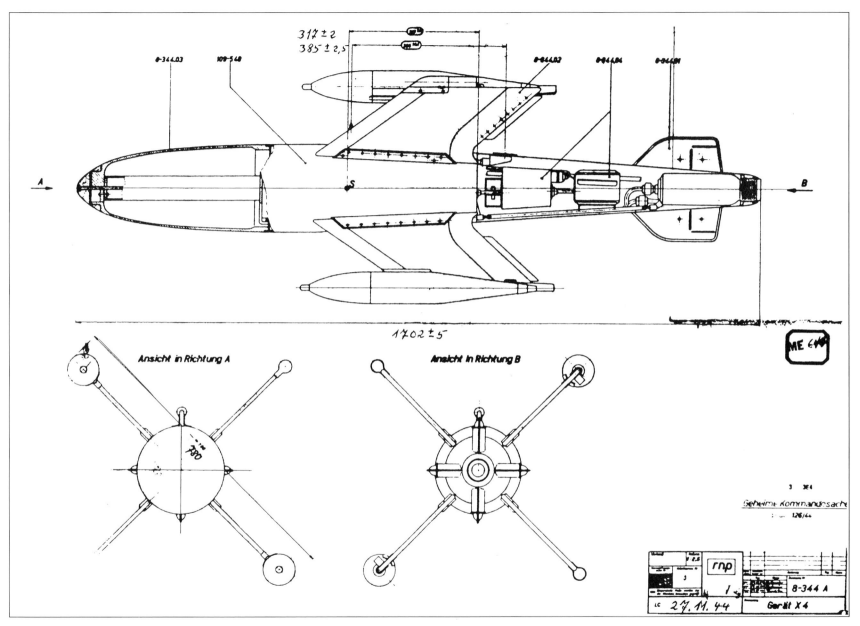

A three-view exterior drawing of the *Luftwaffe X-4* air-to-air missile which was to be the offensive armament of choice for the *Ta 183*. Several were to be hurled into a pack of *B-29s* with the hope that numerous "*superfortress*'" would be knocked out of the sky.

A *MiG 15* scale model as appearing in flight.

The underside of a *MiG 15.*

A *MiG 15's* round turbojet exhaust duct and from this angle its vertical stabilizer is extending up seemingly out of sight.

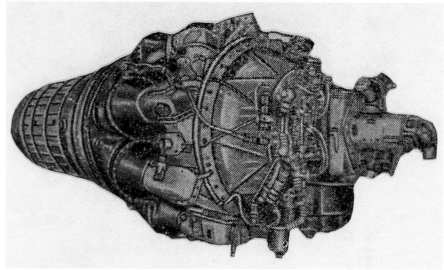

The Soviet "VK-1A" centrifugal turbojet unit...a direct copy of the Rolls-Royce "Nene" and substantially more powerful, too.

A nose starboard side view of a MiG 15. It's bifurcated air intake is readily apparent. From this angle it appears that this machine has three fences per wing but it only had two. It's single air intake duct was split (bifurcated), that is, divided into two ducts so that it could go around the pilot's cockpit seat. The air intake duct on the Pulqui II was bifurcated, too. It does not appear that the Ta 183's air intake duct would have been bifurcated because detailed 3-view drawings from early 1945 show a straight-through air intake duct to its aft HeS 011A.

A MiG 15 handed over to the Americans in Korea by a defecting North Korean airman. From this angle the swept back vertical fin and its horizontal tailplane are clearly visible.

A Korean-era American *F-86A* turbojet powered fighter aircraft giving a good view of its sweptback wings and tailplane.

A rear port side nose view of a *F-86A* turbojet-powered fighter.

Martin Winter (left) and *Hans Multhopp* shown during their employment by the *RAE*-Farnborough, England postwar. *Martin Winter* was *Multhopp's* assistant at *Focke-Wulf* , at *RAE,* and later at *Martin Aircraft.*

RAE-Farnborough, England about mid 1945 or early 1946. *Martin Winter* (first from left back row)* and *Hans Multhopp* (second from left back row).

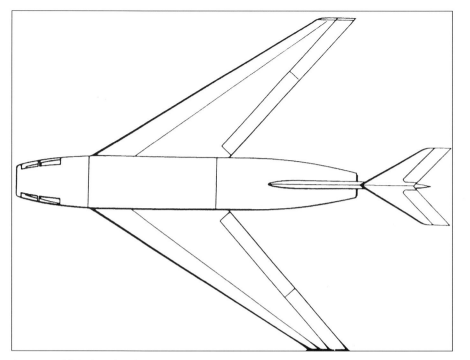

A pen and ink drawing featuring a view from above of *Multhopp*'s 800 mile per hour research aircraft proposed to the *RAE* post war.

A view of the *English Electric's P.1* as seen from below with a complete outline of its swept- back wings. About 1947.

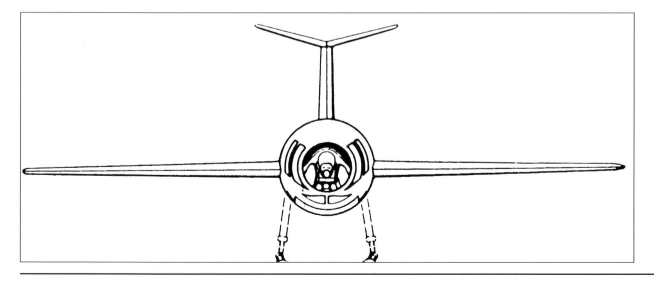

A direct nose on view of the *Multhopp* designed *RAE* research aircraft. The prone pilot position was used successfully by the *Horten* brothers as well as in *DFS' 346* speed of sound machine. *Multhopp's* trade mark appears in its T-tail.

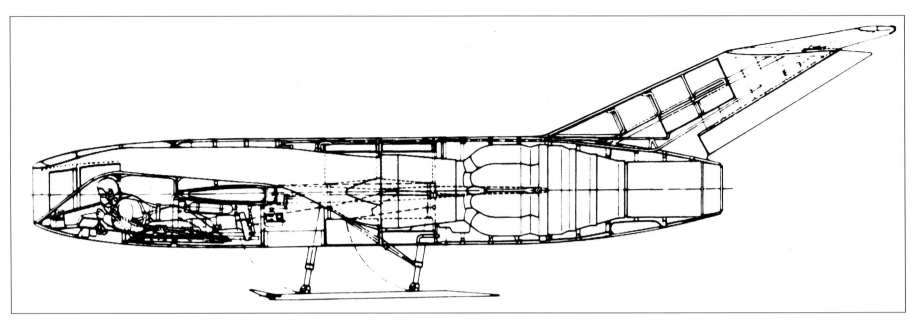

A pen and ink internal illustrated view of the *Multhopp* designed *RAE* research aircraft. This machine was to have been powered by *Rolls-Royce's* first serial production axial flow turbojet engine...the *AJ65* "*Avon*."

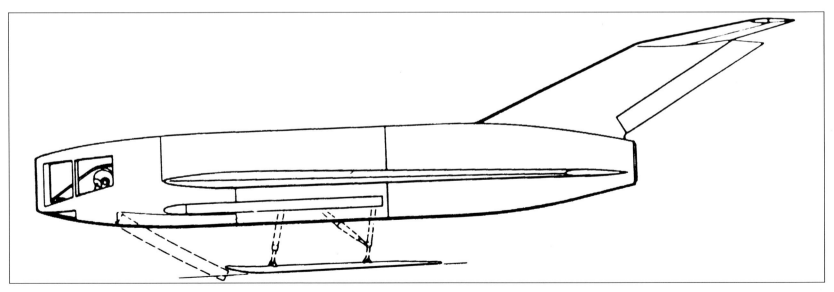

Although *RAE* could not build *Multhopp's* speed of sound design, *English Electric Company* did borrow from the *RAE* machine to build this Mach 2.0 *P.1* "*Lighting*" in the late 1940s. *English Electric* opted to forgo the *Multhopp* T-tail and the prone pilot position in their *P.1*.

A view of the *English Electric's P.1* as seen from above with a view of its port side. About 1947.

The *Multhopp* designed *Martin Aircraft XB-51* in flight in the early 1950s.

The *Multhopp* trademark forever...his signature "*T-tail*" as shown here on the *Martin Aircraft XB-51* in the early 1950s.

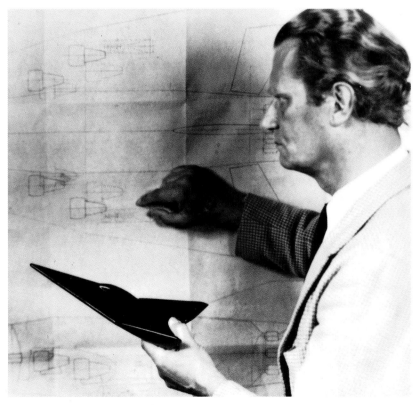

Hans Multhopp holding a scale model of his *Martin Aircraft* space re-entry vehicle known as the *SV*-5 lifting body.

A scale mode of the *Hans Multhopp* designed *SV-5* space re-entry vehicle lifting body by *Martin Aircraft*.

Oh, happy days are here again! Argentine *President Juan Perón,* fond of *Hitler* and *Nazism* in general, welcomes former German aircraft designers such as *Kurt Tank* and *Reimar Horten* in 1946 to the bomber less skies of Argentina. He was especially thrilled to learn that *Tank* had a full set of design drawing for the *183* and with this fighter aircraft *Perón* had dreams of controlling all of South America and kick out the British from Argentina's off shore southern Falkland Island as well.

Perón wanted the *Ta 183* and he believed only the *German* group could produce it...not the capable Frenchmen *Dewoitine*. Here is a *Pulqui II* prototype parked on the tarmac at Córdoba, Argentina in the late 1940s.

Perón had entered the turbojet-powered club before anyone else in South America thanks to *Emilio Dewoitine's IAe 27 Pulqui I*. It was a good machine for its time but *Perón* wanted more. Córdoba, Argentina. 1950.

Below: This is exactly what *Perón* initially wanted from his German aircraft design friends from *Focke-Wulf...Hans Multhopp's Ta 183*. Shown is *Kurt Tank* appearing to describe the scale model *IAe-33's* T-tail to a young child. It's not clear why *Perón* turned to *Tank* rather than *Dewoitine* who was a very capable designer. *Dewoitine* had already given *Perón* a first class fighter in the form of the *IAe 27*...and it is believed that he could perhaps done a better job on the *IAe 33* than *Tank* himself.

A nose-on port side view of *Emilio Dewoitine's Rolls-Royce "Derwent 5"* turbojet-powered *IAe 27 Pulqui I*. It was capable of 528 mph [850 *km/h*] top speed. Not bad for a single engine fighter prototype of the late 1940s.

This is the fighter aircraft which was to be the basis of *Juan Perón's* military offensive...*Multhopp's 183!* However, *Hans Multhopp* declined *Tank's* invitation to join him and his former colleagues in Argentina. Scale model by *Günter Sengfelder*.

A wooden scale model of *Kurt Tank's Pulqui II* (*IAe 33*) suspended in a wind tunnel at the *Instituto Aerotécnico* (*IAe*)-Córdoba, Argentina. About 1948.

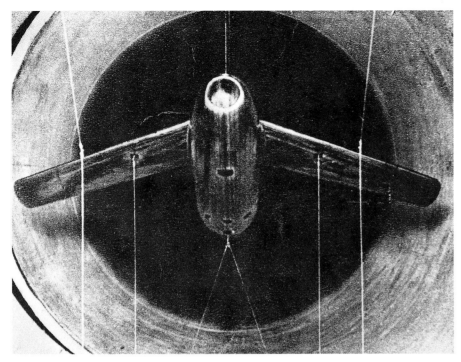

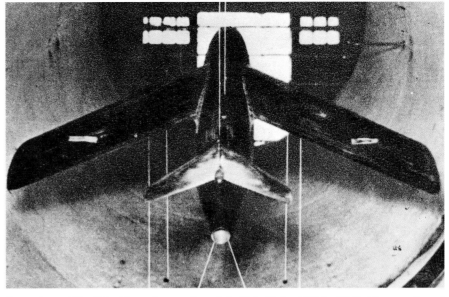

Rear view of *Kurt Tank's* scale wood model of his *Pulqui II* (*IAe 33*) fighter version of the *Ta 183* in the Argentine *Instituto Aerotécnico* (*IAe*) wind tunnel at Córdoba. About 1948.

Nose on view of *Kurt Tank's* scale wood model of his *Pulqui II* (*IAe 33*) fighter version of the *Ta 183* in the Argentine *Instituto Aerotécnico* (*IAe*) wind tunnel at Córdoba. About 1948.

A poor quality photo of one of the five *Pulqui II's* (*IAe 33*) under construction at the *Instituto Aerotécnico* (*IAe*), Córdoba, Argentina. About 1950.

Reimar and *Gisela Horten* at the time of their marriage at Córdoba, Argentina in *1949. Gisela Horten* held the unofficial South American record for altitude among female sailplaners. In 1945 she had reached 9,842 feet [3,000 meters]. At the same time she had set a female endurance record of five hours.

Kurt Tank's Pulqui II always drew a crowd of admirers during its test trials.

A nose port side view of a *Pulqui II* as it waits for takeoff clearance at Córdoba, Argentina during its flight-testing trials in the late 1940s. It's painting scheme was white overall except for the red paint around its air intake and stripe extending back along the fuselage. The word "*Pulqui II*" in red began mid fuselage beneath the wing root's leading edge on both port and starboard sides.

A *Pulqui II* being readied for a test flight by *Major Otto Behrens*. A farmer tractor served as a tow vehicle for the relatively lightweight fighter aircraft prototype.

A very determined-looking *Kurt Tank* in the cockpit of his *Pulqui II*. He was facing the possibility that the machine was, indeed, appearing to be a bonafide design failure.

Kurt Tank appears to be giving a typical right arm extended *NAZI* salute to the crowd as he disembarks from a test flight with the *Pulqui II*. Perhaps *Tank* out of habit momentarily forgot that he was in Argentina not *NAZI* Germany and the year was 1950 and not the early to mid 1940s. In the far right of the photo another arm appears to be raised in a return salute.

Top Left: Kurt Tank back on the ground after a test flight. The tailplane of the *Pulqui II* is seen just over *Tank's* head. The man to the far right in the photo is *Otto Pabst*, noted gas dynamics expert from the former *Focke-Wulf Flugzeugbau*-Bremen. *Top Center: Kurt Tank* (left) and *President Juan Perón* celebrating happy times over the initial testing of the *Pulqui II*. It is not entirely clear when *Perón* turned sour over the *Pulqui II* and *Tank* as well. But he did. *Perón* fired *Tank* and all sixty of his former employees. In an ironic twist of fate *Martin Aircraft* sent *Multhopp* down to Argentina to offer employment to a few of his former colleagues. It was as if *Multhopp* was just waiting to pick up the best of the ex-*Focke-Wulf* engineers knowing somehow that *Tank's* version of his *183* was going to be a spectacular failure. *Reimar Horten* and *Hans Multhopp* corresponded with each other on a regular basis (they had been friendly back in Germany during the war) and the two pretty much knew from the day *Tank* announced that he was going to place the *183's* wing shoulder-high that this *Pulqui II* wasn't meant to be. *Top Right: Kurt Tank* (right) and *Senor Nuestro*, from the *Instituto Aerotécnico* and project director on the *Pulqui II* program. Late 1940s.

A *Pulqui II* seen from below giving a good overall view of its shoulder-high wing position.

A front starboard side view of the *Pulqui II*. This official photo shows a nice view of its *Multhopp* T-tail as the machine hunches down on its main landing gear.

A nose port side view of the Pulqui II's twin *Oberlikon* 20 mm gun ports.

Argentine *Captain Manuel* out in front of a *Pulque II* and with a good view of its air intake. *Manuel* would later die while test flying *Tank's* version of *Multhopp's* design.

Otto Behrens beginning his takeoff run in a *Pulque II*...his last.

Above: This *Pulqui II*, on outdoor display at Buenos Aires' *Aero Parque*, has a red paint strip which is non regulation nor was the word "*Pulqui II*" positioned as far forward as it is now on this last survivor. Normally each of the five *Pulqui II* proto-type*s* had their red fuselage strip proceed aft from the air intake and narrow down as it reached the tail area. Photograph by the author in the mid 1980s.

One of *E-Stelle* Rechlin's finest test pilots: *Otto Behrens* stands on the starboard side of a *Pulqui II*. He would later die while test flying the "brute" as he described it.

A nose-on view of the last surviving *Pulqui II* prototype out of the five constructed in the late 1950s. Note the nonregulation paint job. Photographed at the *Aero Parque*, Buenos Aires, Argentina by the author in the mid 1980s.

A full port side view of the *Pulqui II* at the *Aero Parque*. The double rectangular items on the aft fuselage are speed brakes in their full retracted position. Photograph by the author in the mid 1980s.

A close up of the *Pulqui II's* port wing tip including its control surfaces. Photograph by the author in the mid 1980s.

A good view of the last surviving *Pulqui II's* port upper surface wing root. Photograph by the author in the mid 1980s.

A close up view of the *Pulqui II's* swept back vertical fin root at the extreme end of the fuselage. Shown, too, are the port side speed brakes in their full-closed position and the short, small fin each side of the fuselage. Photograph by the author in the mid 1980s.

A close up view of the *Pulqui II's* upper surface port side wing root. Photograph by the author in the mid 1980s.

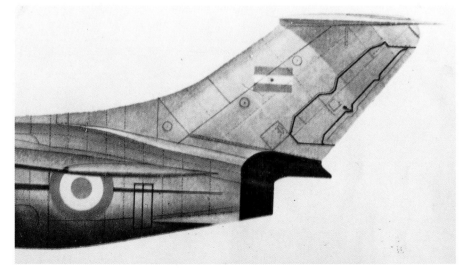

A pen and ink illustration of the *Pulqui II's* vertical fin position on the dorsal fuselage.

A view of the *Pulqui II's* turbojet engine exhaust duct and the two small fins extending forward for a short distance before blending in to the fuselage outer surface. Photograph by the author in the mid 1980s.

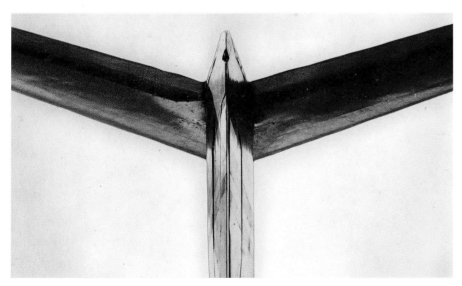

A dramatic view of the vertical fin and horizontal tailplane of a *Pulqui II* as seen from the rear. Photograph by the author in the mid 1980s.

The turbojet exhaust end of the "*Pulqui II*" on permanent outdoor display at the *Aero Parque* showing its characteristic *Multhopp* T-tail. Notice, too, how *Kurt Tank* repositioned the wing's from *Multhopp's* mid fuselage location on his *Ta 183* to a shoulder high location on the *Pulqui II*...a considerable height beyond the recommendations of *Multhopp* as well as the *MiG* team. Photograph by the author in the mid 1980s.

The horizontal tailplane atop the swept-back vertical stabilizer on the *Pulqui II*. Photograph by the author in the mid 1980s.

Kurt Tank, second from left, along with *President Juan Perón* and senior Argentine Air Force officers. In the back ground is the tailplane of a *Pulqui II*. This is interesting. The elevators are in a up position while the rudder, too, is turned to starboard. The photo is from the early 1950s.

The starboard side of the *Pulqui II* prototype. Photograph by the author in the mid 1980s.

President Juan Perón (in the white hat) standing on a *Pulqui II* prototype's starboard wing inspecting the cockpit. In the foreground is the backside of the starboard landing flap showing its stifling rib/pieces along its length.

A full port side view of the *Pulqui II's* vertical fin with its attached rudder. Photograph by the author in the mid 1980s.

The starboard side of the *Pulqui II* prototype at the *Aero Parque*. Note the wing fence on the upper surface of the wing. At the time of the flight testing *Tank* and his associates were seeking ways to reduce the "superstall" which now was hampering the flight testing. Some historians claim that the problem was rectified but it wasn't limiting construction to five prototypes before the entire *IAe 33* program was killed and *Tank* and his group fired. Photograph by the author in the mid 1980s.

A rear starboard side view of the *Pulqui II* prototype at the *Aero Parque*. Notice that its starboard side speed brake is partially extended out of the fuselage. A tin cover closes up the exhaust duct. Photograph by the author in the mid 1980's.

The *Pulqui II's* aft fuselage-mounted speed brake partially extended. Photograph by the author in the mid 1980s.

The *Pulqui II's* starboard side speed brake in its full open position. Photograph by the author in the mid 1980s.

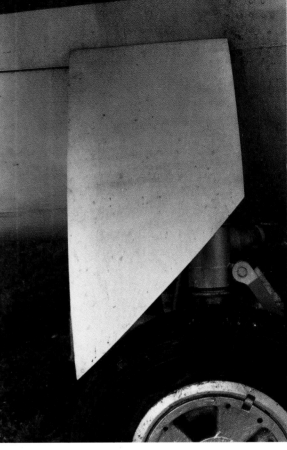

The starboard side main landing gear cover door on all the *Pulqui II* prototypes. Photograph by the author in the mid 1980s.

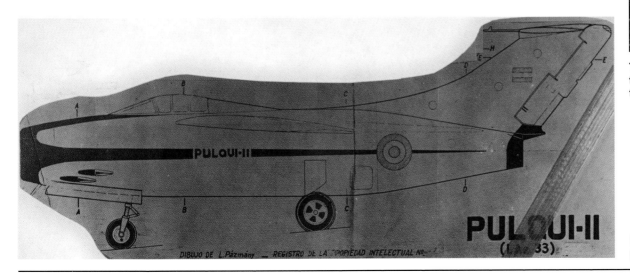

Port side view of a *Pulqui II* fuselage showing sectional profile locations for building a scale model.

Argentine test pilot *Captain Manuel* is shown out front of a *Pulqui II*. Note the bifurcated air intake...similar to the *MiG 15*. The bifurcated air intake duct may not have been used on the *Ta 183*.

A cutaway "Nene's" port side showing its frontal gear box, compressor wheel, combustion cans, turbine "hot" wheel, and jet pipe.

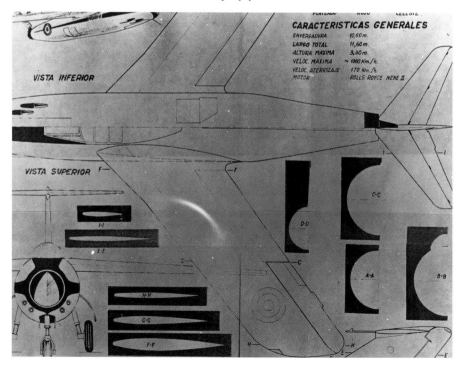

Profile sections of a *Pulqui II's* fuselage, wing, and tailplane for the scale model builder.

The starboard side view of a *Rolls-Royce* "*Nene*" centrifugal turbojet engine with its screened air intake opening. The "*Nene*" powered all five *Pulqui II* prototypes.

Two *Ta 183s* running a full throttle with their jet pipe flame burning bright in hot pursuit of *B-29s*. Digital image by *Mario Merino*.

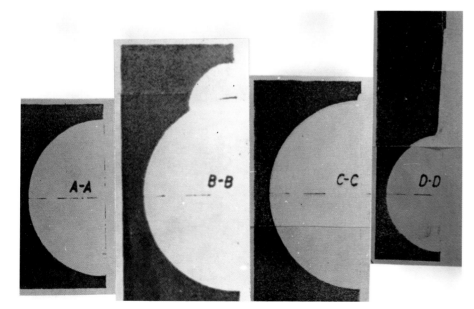

Fuselage sectional profiles for a *Pulqui II*.

Wing section profiles for a *Pulqui II*.

A complete view of the *Pulqui II's* starboard main wheel, oleo strut, and wheel-well cover door. Photograph by the author in the mid 1980s.

A nose on ground view of the *Pulqui II's* tricycle landing gear especially its nose wheel. The nose wheel has a wide groove in its center, perhaps a 1/2 inch lower than the adjacent rubber. Photograph by the author in the mid 1980s.